Contents

Preface..2

Executive Summary...3

About the Author..4

Chapter 1: Fabaceae..5-24

Chapter 2: Herbarium...25-33

Preface

This book consists of two chapter. The first chapter gives fabaceae general charater and its important to human life. The second chapter gives fabaceae family herbarium like Mimosa pudica, Cajanus cajan, Indigofera hirsute etc.

The authors are thankful to all the friends for their continue support and encouragement. If you discover any error in this book please notify me anupamrajak1234@gmail.com

Executive Summary

This book is usefull to students, schlors, reasearchers, scientists and interested family. The Fabaceae family is economically and ecologically very important family. This family are widely distributed in the world. This families are provide us timbers, dyes and food. The Fabaceae family are medicinally important plants.

About the Author

Anupam Rajak received his B.Sc in Botany from the Raghunathpur College, Sidho-Kanho-Birsha University. He has published several articles in international journal. His email address is anupamrajak1234@gmail.com

Chapter 1

Fabaceae

Fabaceae is also called leguminosae family. The type genus of fabaceae family is faba. Fabaceae is third largest family in angiosperms. Fabaaceae family comprises of 20,000 species in worldwide in distribution. They are medicinally important plants. Fabaaceae family are cosmopolitan in distribution. Fabaceae family are 630 genera.

Figure 1. Fabaceae Family Crotolaria sp. (Photo Credit: Anupam Rajak)

Scientific Classification:

Kingdom	Plantae
Divison	Magnoliophyta
Class	Magnoliopsida
Order	Fabales
Family	Fabaceae

Characters:

1. Habit: Fabaceae families are herb, shrub, trees or vines. Some families are tendril or climbers. They are annual or perennial.

2. Leaves: Leaves are simple, usally alternate and compound. Leaves are pinnately or palmately compound.

Figure 2. Fabaceae Family leaves (Photo Credit: Anupam Rajak)

3. Flowers: Flowers are usally bilaterally symmetrical, actinomorphic (Mimosoideae), zygomorphic and bisexual.

Figure 3. Fabaceae family flowers (Photo Credit: Pixabay)

4. Inflorescence: Inflorescence are of racemose or cymose type.

Figure 4. Fabaceae Family (Photo Credit: Anupam Rajak)

5. Calyx: Calyx are usally green in colour. Number of the calyx are 5 or sometimes 4.

6. Corolla: Corolla are free or united. Number of the corolla are 5 or sometimes 4.

7. Androecium: Stamens are usally 10. Stamens are monodelphous or diadelphous and free.

8. Gynoecium: Carpels are free. Carpels are monocarpellary.

9. Fruit: Legume or pod.

Figure 5. Fabaceae family fruit (Photo Credit: Anupam Rajak)

10. Roots: Roots are tap roots, fibrous or adventitious. Root nodules are found on the fabaceae family.

Economic Importance of Fabaceae:

Fabaceae family are ecologically and economically important plants. These families provide us to food, timbers, vegetables, woods, and dyes.

Pulses are rich in protein such as Pisum sativum, Vicia faba, Vigna mungo etc.

Soyabean and ground nut are used to extract oil. Fibers are extract from sun hemp and sesbania cannabina.

The family leguminosae or fabaceae family is divided into 3 subfamilies_

i. Papilionaceae or papilionoideae.

ii. Caesalpiniodeae.

iii. Mimosoideae.

Papilionaceae:

The members of the papilionaceae family are distributed worldwide. It is the largest family under the order leguminales. Its includes 375 genera.

Habit: Plants are herbs, shrubs and tree.

Roots: Roots may be tap root or branched. Root nodules are found.

Stem: Erect or Climbing.

Leaves: Leaves may be alternate, pinnate or whorled.

Inflorescence: Raceme or spike.

Flowers: Zygomorphic.

Some Pictures of Fabaceae family are-

Acacia

Acacia is a tree belonging to the family Fabaceae. Many people eat acacia seeds. Gums are extracted from acacia.

Figure 6. Acacia Sp. (Photo Credit: Anupam Rajak)

Pea:

Pea is plant. Pods can be green or yellow. We eats pea seeds.

Figure 7. Pisum Sativum (Photo Credit: Anupam Rajak)

Figure 8. Butea monospermae (Photo Credit: Anupam Rajak)

Figure 9. Clitoria ternata (Photo Credit: Anupam Rajak)

Figure 10. Lablab purpureus (Photo Credit: Anupam Rajak)

Figure 10. Crotolaria retusa (Photo Credit: Anupam Rajak)

Figure 11. Pigeon pea (Photo Credit: Anupam Rajak)

Figure 12. Acacia auriculiformis (Photo Credit: Anupam Rajak)

Figure 13. Phaseolus sp. (Bean) (Photo Credit: Anupam Rajak)

Figure 14. Fabaceae family (Photo Credit: Anupam Rajak)

Figure 15. How local people grows pigeon pea (Photo Credit: Anupam Rajak)

Chapter 2

Herbarium

All the plants are collected from Bolpur-Santiniketan area. Bolpur –Santiniketan are floristicscally very rich. After collection, plants are pressed. Then some-day plants are dried. After dry, plants are attached with the herbarium sheet.

Mimosa Pudica:

Mimosa Pudica is annual or perennial herb.

Figure 1. Mimosa pudica (Photo Credit: Anupam Rajak)

Figure 2. Cajanus Cajan (Photo Credit: Anupam Rajak)

Figure 3. Indigofera hirsuta (Photo Credit: Anupam Rajak)

Figure 4. Cicer arietinum (Photo Credit: Anupam Rajak)

Figure 5. Crotolaria juncea (Photo Credit: Anupam Rajak)

Figure 6. Tephrosia sp. (Photo Credit: Anupam Rajak)

Figure 7. Crotolaria retusa (Photo Credit: Anupam Rajak)

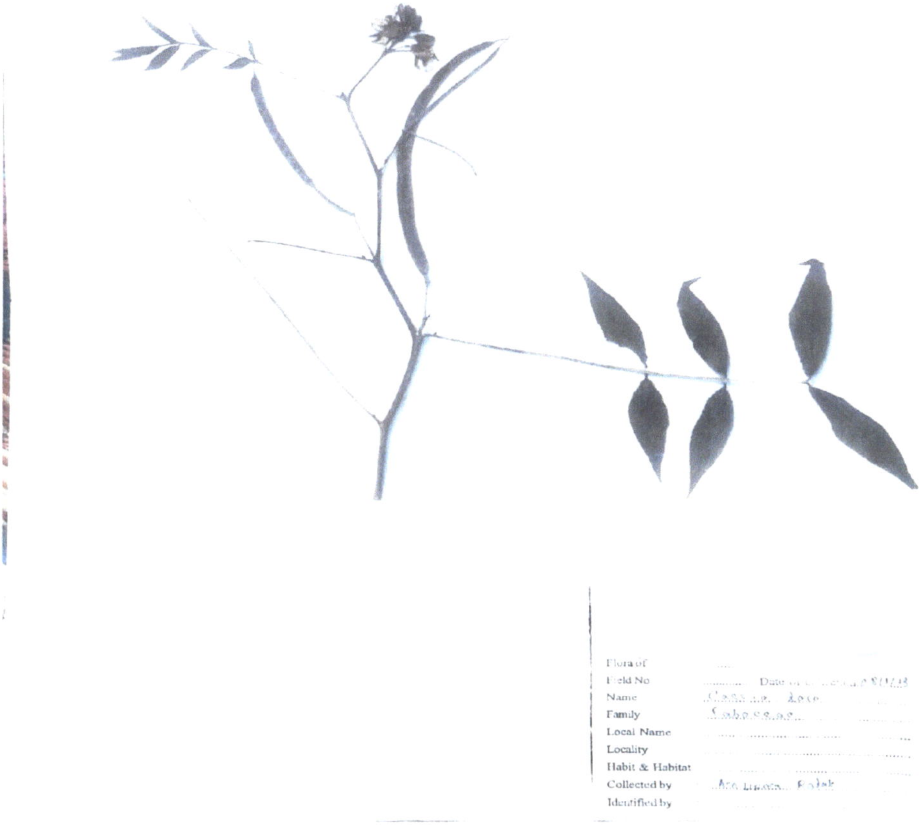

Figure 8. Cassia sp. (Photo Credit: Anupam Rajak)

References:

1. Doyle, J.J., J.A. Chappill, C.D. Bailey, and T. Kajita. 2000. Towards a comprehensive phylogeny of legumes: evidence from rbcL sequences and non-molecular data. Pages 1-20 in Advances in legume systematics, part 9, (P. S. Herendeen and A. Bruneau, eds.). Royal Botanic Gardens, Kew, UK.

2. Gepts, P., W.D. Beavis, E.C. Brummer, R.C. Shoemaker, H.T. Stalker, N.F. Weeden, and N.D. Young. 2005. Legumes as a model plant family. Genomics for food and feed report of the cross-legume advances through genomics conference. Plant Physiol. 137: 1228 – 1235.

3. Lavin, M., P.S. Herendeen, and M.F. Wojciechowski. 2005. Evolutionary rates analysis of Leguminosae implicates a rapid diversification of lineages during the Tertiary. Syst. Biol. 54: 530-549.

4. Crepet, W. L., and D. W. Taylor. 1985. The diversification of the Leguminosae: first fossil evidence of the Mimosoideae and Papilionoideae. Science 288: 1087-1089.

5. Crepet, W. L., and D. W. Taylor. 1986. Primitive mimosoid flowers from the Paleocene-Eocene and their systematic and evolutionary implications. American J. Botany 73: 548-563.

6. Crepet, W. L., and P. S. Herendeen. 1992. Papilionoid flowers from the early Eocene of southeastern North America. Pages 43–55 in Advances in Legume Systematics, part 4, the fossil record (P. S. Herendeen and D. L. Dilcher, eds.). Royal Botanic Gardens, Kew, UK.

7. Allan, G. J., and J. M. Porter. 2000. Tribal delimitation and phylogenetic relationships of Loteae and Coronilleae (Faboideae: Fabaceae) with special reference to Lotus: evidence from nuclear ribosomal ITS sequences. American J. Botany 87: 1871-1881.

8. Angiosperm Phylogeny Group [APG]. 2003. An update of the Angiosperm Phylogeny Group classification for the orders and families of flowering plants: APG II. Botanical J. Linnean Society 141: 399-436.

9. Bruneau, A., F. Forest, P. S. Herendeen, B. B. Klitgaard, and G. P. Lewis. 2001. Phylogenetic relationships in the Caesalpinioideae (Leguminosae) as inferred from chloroplast trnL intron sequences. Systematic Botany 26: 487-514.

10. Chappill, J. A. 1995. Cladistic analysis of the Leguminosae: the development of an explicit hypothesis. Pages 1-10 in Advances in Legume Systematics, part 7, phylogeny (M. D. Crisp and J. J. Doyle, eds.). Royal Botanic Gardens, Kew, UK.

11. Doyle, J.J., J.L. Doyle, J.A. Ballenger, E.E. Dickson, T. Kajita, and H. Ohashi. 1997. A phylogeny of the chloroplast gene rbcL in the Leguminosae: taxonomic correlations and insights into the evolution of nodulation. Amer. J. Bot. 84: 541-554.

12. Pennington, R.T., M. Lavin, H. Ireland, B.B. Klitgaard, and J. Preston. 2001. Phylogenetic relationships of basal papilionoid legumes based upon sequences of the chloroplast trnL intron. Syst. Bot. 26: 537-566.

13. Chase, M. W., D. E. Soltis, R. G. Olmstead, D. Morgan, D. H. Les, B. D. Mishler, M. R. Duvall, R. A. Price, H. G. Hills, Y.-L. Qiu, K. A. Kron, J. H. Rettig, E. Conti, J. D. Palmer, J. R. Manhart, K. J. Sytsma, H. J. Michaels, W. J. Kress, K. G. Karol, W. D. Clark, M. Hedrén, B. S. Gaut, R. K. Jansen, K.-J. Kim, C. F. Wimpee, J. F. Smith, G. R. Furnier, S. H. Strauss, Q.-Y. Xiang, G. M. Plunkett, P. S. Soltis, S. M. Swensen, S. E. Williams, P. A. Gadek, C. J. Quinn, L. E. Eguiarte, E. Golenberg, G. H. Learn Jr., S. W. Graham, S. C. H. Barrett, S. Dayanandan, and V. A. Albert. 1993. Phylogenetics of seed plants: An analysis of nucleotide sequences from the plastid gene rbcL. Annals of the Missouri Botanic Garden 80: 528-580.

14. Herendeen, P. S., and D. L. Dilcher. 1990. Diplotropis (Leguminosae, Papilionoideae) from the Middle Eocene of southeastern North America. Systematic Botany 15: 526-533.

15. Herendeen, P. S., and D. L. Dilcher. 1991. Caesalpinia subgenus Mezoneuron (Leguminosae, Caesalpinioideae) from the Tertiary of North America. American J. Botany 78: 1-12.

16. Herendeen, P. S., and S. Wing. 2001. Papilionoid legume fruits and leaves from the Paleocene of northwestern Wyoming. Botany 2001 Abstracts, published by Botanical Society of America (http://www.botany2001.org/).

17. Herendeen, P. S., A. Bruneau, and G. P. Lewis. 2003. Phylogenetic relationships in caesalpinioid legumes: a preliminary analysis based on morphological and molecular data. Pages 37-62 in Advances in Legume Systematics, part 10, higher level systematics (B.B. Klitgaard and A. Bruneau, eds.). Royal Botanic Gardens, Kew, UK.

18. Hu, J.-M., M. Lavin, M. F. Wojciechowski, and M.J. Sanderson. 2000. Phylogenetic systematics of the tribe Millettieae (Leguminosae) based on matK sequences, and implications for evolutionary patterns in Papilionoideae. American J. Botany 87: 418-430.

19. Kajita, T., H. Ohashi, Y. Tateishi, C. D. Bailey, and J. J. Doyle. 2001. rbcL and legume phylogeny, with particular reference to Phaseoleae, Millettieae, and allies. Systematic Botany 26: 515-536.

20. Käss, E., and M. Wink. 1996. Molecular evolution of the Leguminosae: phylogeny of the three subfamilies based on rbcL sequences. Biochemical Systematics and Evolution 24: 365-378.

21. Käss, E., and M. Wink. 1997. Phylogenetic relationships in the Papilionoideae (Family Leguminosae) based on nucleotide sequences of cpDNA (rbcL) and ncDNA (ITS1 and 2). Molecular Phylogenetics and Evolution 8:65-88.

22. Lavin, M., J. J. Doyle, and J. D. Palmer. 1990. Evolutionary significance of the loss of the chloroplast--DNA inverted repeat in the Leguminosae subfamily Papilionoideae. Evolution 44: 390-402.

23. Lavin. M., R. T. Pennington, B. B. Klitgaard, J. I. Sprent, H. C. de Lima, and P. E. Grasson. 2001. The Dalbergioid legumes (Fabaceae): delimitation of a pantropical monophyletic clade. American J. Botany 88: 503-533.

24. Lavin, M., M. F. Wojciechowski, P. Gasson, C. E. Hughes, and E. Wheeler. 2003. Phylogeny of robinioid legumes (Fabaceae) revisited: Coursetia and Gliricidia recircumscribed, and a biogeographical appraisal of the Caribbean endemics. Systematic Botany 28: 387–409.

25. Lavin, M., P. S. Herendeen, and M. F. Wojciechowski. 2005. Evolutionary rates analysis of Leguminosae implicates a rapid diversification of lineages during the Tertiary. Systematic Biology 54: 530-549.

26. Lewis, G., B. Schrire, B. MacKinder, and M. Lock (eds). 2005. Legumes of the world. Royal Botanical Gardens, Kew, UK.

27. Polhill, R.M. 1994. Classification of the Leguminosae. Pages xxxv–xlviii in Phytochemical dictionary of the Leguminosae (F. A. Bisby, J. Buckingham, and J. B. Harborne, eds.). Chapman and Hall, New York, NY.

28. Wojciechowski, M. F., M. Lavin, and M. J. Sanderson. 2004. A phylogeny of legumes (Leguminosae) based on analysis of the plastid matK gene resolves many well-supported subclades within the family. American Journal of Botany 91: 1846-1862.

29. Crisp, M. D., S. Gilmore, and B-E. Van Wyk. 2000. Molecular phylogeny of the genistoid tribes of papilionoid legumes. Pages 249-276 in Advances in Legume Systematics, part 9 (P. S. Herendeen and A. Bruneau, eds.). Royal Botanic Garden, Kew, UK.

30. Dickison, W. C. 1981. The evolutionary relationships of the Leguminosae. Pages 35-54 in Advances in Legumes Systematics, part 1 (R. M. Polhill and P. H. Raven, eds.) Royal Botanic Gardens, Kew, UK.

31. Dilcher, D. L., P. S. Herendeen, and F. Hueber. 1992. Pages 33-42 in Advances in Legume Systematics, part 4, the fossil record (P. S. Herendeen and D. L. Dilcher, eds.). Royal Botanic Gardens, Kew, UK.

32. Doyle, J.J., J.L. Doyle, J.A. Ballenger, E.E. Dickson, T. Kajita, and H. Ohashi. 1997. A phylogeny of the chloroplast gene rbcL in the Leguminosae: taxonomic correlations and insights into the evolution of nodulation. American J. Botany 84: 541-554.

33. Doyle, J. J., J. A. Chappill, C. D. Bailey, and T. Kajita. 2000. Towards a comprehensive phylogeny of legumes: evidence from rbcL sequences and non-molecular data. Pages 1 -20 in Advances in legume systematics, part 9, (P. S. Herendeen and A. Bruneau, eds.). Royal Botanic Gardens, Kew, UK.

34. Gepts, P., W. D. Beavis, E. C. Brummer, R. C. Shoemaker, H. T. Stalker, N. F. Weeden, and N. D. Young. 2005. Legumes as a model plant family. Genomics for food and feed report of the cross-legume advances through genomics conference. Plant Physiology 137: 1228–1235.

35. Graham, P. H., and C. P. Vance. 2003. Legumes: importance and constraints to greater use. Plant Physiol. 131: 872 – 877.

36. Herendeen, P. S. 1992. The fossil history of Leguminosae from the Eocene of southeastern North America. Pages 85-160 in Advances in Legume Systematics, part 4, the fossil record (Herendeen, P. S., and D. L. Dilcher, eds.). Royal Botanic Gardens, Kew, UK.

37. Herendeen, P. S. 2001. The fossil record of the Leguminosae: recent advances. In Legumes Down Under: the Fourth International Legume conference, Abstracts, 34–35. Australian National University, Canberra, Australia.

38. Herendeen, P. S., W. L. Crepet, and D. L. Dilcher. 1992. The fossil history of the Leguminosae: phylogenetic and biogeographic implications. Pages 303 – 316 in Advances in Legume Systematics, part 4, the fossil record (P. S. Herendeen and D .L. Dilcher, eds). Royal Botanic Gardens, Kew, UK.

39. Luckow, M., J. T. Miller, D. J. Murphy, and T. Livshultz. 2003. A phylogenetic analysis of the Mimosoideae (Leguminosae) based on chloroplast DNA sequence data. Pages 197-220 in Advances in Legumes Systematics, part 10, higher level systematics (B. B. Klitgaard and A. Bruneau, eds.). Royal Botanic Gardens, Kew, UK.

40. McKey, D. 1994. Legumes and nitrogen: the evolutionary ecology of a nitrogen-demanding lifestyle. Pages 211–228 in Advances in Legume Systematics, part 5, the nitrogen factor (J. I. Sprent and D. McKey, eds.). Royal Botanic Gardens, Kew, UK.

41. Manos, P. S., and A. M. Standford. 2001. The biogeography of Fagaceae: tracking the Tertiary history of temperate and subtropical forests of the Northern Hemisphere. International J. of Plant Sciences 162: S77-S93.

42. Morley, R. J. 2000. Origin and Evolution of Tropical Rain Forests. John Wiley & Sons. Pp. 378.

43. Müller-Stoll W. R. and E. Mädel. 1967. Die fossilen Leguminosen-Hölzer. Eine revision der mit Leguminosen verglichenen fossilen Hölzer und Beschreibungen älterer und neuer Arten. Palaeontographica, Abt. B, 119: 95-174.

44. Pennington, R. T., M. Lavin, H. Ireland, B. B. Klitgaard, and J. Preston. 2001. Phylogenetic relationships of basal papilionoid legumes based upon sequences of the chloroplast trnL intron. Systematic Botany 26: 537-566.

45. Persson, C. 2001. Phylogenetic relationships in Polygalaceae based on plastid DNA sequences from the trnL-F region. Taxon 50: 763-779.

46. Polhill, R. M. 1994. Classification of the Leguminosae. Pages xxxv–xlviii in Phytochemical Dictionary of the Leguminosae (F. A. Bisby, J. Buckingham, and J. B. Harborne, eds.). Chapman and Hall, New York, NY.

47. Polhill, R. M., and P. H. Raven (eds.). 1981. Advances in legume systematics, parts 1 and 2. Royal Botanic Gardens, Kew, UK.

48. Wojciechowski, M. F., M. Lavin, and M. J. Sanderson. 2004. A phylogeny of legumes (Leguminosae) based on analysis of the plastid matK gene resolves many well-supported subclades within the family. American J. Botany 91: 1846-1862.

49. Wojciechowski, M. F., M. J. Sanderson, K. P. Steele, and A. Liston. 2000. Molecular phylogeny of the "temperate herbaceous tribes" of papilionoid legumes: a supertree approach. Pages 277-298 in Advances in Legume Systematics, part 9 (P. S. Herendeen and A. Bruneau, eds.). Royal Botanic Gardens, Kew, UK.0

50. Xu Z., Deng M. (2017) Fabaceae or Leguminosae. In: Identification and Control of Common Weeds: Volume 2. Springer, Dordrecht

51. Klitgård, B.B. & Lewis, G.P. (2010). Neotropical Leguminosae (Mimosoideae). In: Milliken, W., Klitgård, B. & Baracat, A. (2009 onwards), Neotropikey - Interactive key and information resources for flowering plants of the Neotropics.http://www.kew.org/science/tropamerica/neotropikey/families/Leguminosae_(Mimosoideae).htm.

52. Zarucchi, J.L. 1993. Fabaceae. In: L. Brako & J.L. Zarucchi (eds.). Catalogue of the flowering plants and gymnosperms of Peru. Monographs in Systematic Botany from the Missouri Botanical Garden, vol. 45: Fabaceae pp. 444-527.

53. Clement, B.A., Goff, C.M., Forbes, T.D.A. Toxic Amines and Alkaloids from Acacia rigidula, Phytochem. 1998, 49(5), 1377.

54. Wojciechowski, Martin F., Johanna Mahn, and Bruce Jones. 2006. Fabaceae. legumes. Version 14 June 2006. The Tree of Life Web Project, http://tolweb.org/Pan2010 (Journal) : Pan, A. D et al (2010),[{{{url}}} Detarieae sensu lato (Fabaceae) from the Late Oligocene (27.23 Ma) Guang River flora of north-western Ethiopia.], Botanical Journal of the Linnean Society:163(1):44. doi=10.1111/j.1095-8339.2010.01044.x

55. Polhill, R. M., P. H. Raven, and C. H. Stirton. 1981. Evolution and systematics of the Leguminosae. Pages 1-26 in Advances in Legume Systematics, part 1 (R. M. Polhill and P. H. Raven, eds.). Royal Botanic Gardens, Kew, UK.

56. Raven, P. H. and D. I. Axelrod. 1974. Angiosperm biogeography and past continental movements. Annals of the Missouri Botanic Garden 61: 539-657.

57. Rundel, P. W. 1989. Ecological success in relation to plant form and function in the woody legumes. In C.H. Stirton and J.L. Zarucchi (eds.). Advances in legume biology, Monographs in Systematic Botany from the Missouri Botanical Gardens 29: 377-398.

58. Sanderson, M. J., and M. F. Wojciechowski. 1996. Diversification rates in a temperate legume clade: are there "so many species" of Astragalus (Fabaceae)? American J. Botany 83: 1488-1502.

59. Savolainen, V., M. W. Chase, S. B. Hoot, C. M. Morton, D. E. Soltis, C. Bayer, M. F. Fay, A. Y. de Bruijn, S. Sulllivan, and Y.-L. Qiu. 2000. Phylogenetics of flowering plants based on combined analysis of plastid atpB and rbcL sequences. Systematic Biology 49: 306-362.

60. Soltis, D.E., P.S. Soltis, D.R. Morgan, S.M. Swensen, B.C. Mullin, J.M. Dowd, and P.G. Martin. 1995. Chloroplast gene sequence data suggest a single origin of the predisposition for symbiotic nitrogen fixation in angiosperms. Proceedings of the National Academy of Sciences USA 92: 2647-2651.

61. Soltis, D. E., P. S. Soltis, M. W. Chase, M. E. Mort, D. C. Albach, M. Zanis, V. Savolainen, W. H. Hahn, S. B. Hoot, M. F. Fay, M. Axtell, S. M. Swenson, L. M. Prince, W. J. Kress, K. C. Nixon, and J. S. Farris. 2000. Angiosperm phylogeny inferred from 18S rDNA, rbcL, and atpB sequences. Botanical J. Linnean Society 133: 381-461.

62. Sprent, J. I. 2001. Nodulation in legumes. Royal Botanic Gardens, Kew, UK.

63. Steele, K. P., E. Tizon, R. C. Evan, C. S. Campbell, and M. F. Wojciechowski. 2000. Sister group relationships of Fabaceae and Rosaceae: phylogenetic relationships of Eurosids I. American J. Botnay 87: S160 (abstract).

64. Thorne, R. F. 1992. Classification and geography of the flowering plants. Botanical Review 58: 225-348.

65. Wikström, N., V. Savolainen, and M. W. Chase. 2001. Evolution of the angiosperms: calibrating the family tree. Proceedings of the Royal Society of London, Series B 268: 2211-2220.

66. Wing, S. L., F. Herrera, and C. Jaramillo. 2004. A Paleocene flora from the Cerrajón Formation, Guajíra Peninsula, northeastern Colombia. Pages 146-147 in VII International Organization of Paleobotany Conference Abstracts (21-26 March). Museo Egidio Feruglio, Trelew, Argentina.

67. Wojciechowski, M. F. 2003. Reconstructing the phylogeny of legumes (Leguminosae): an early 21st century perspective. Pages 5-35 in Advances in Legume Systematics, part 10, higher level systematics (B. B. Klitgaard and A. Bruneau, eds.). Royal Botanic Gardens, Kew, UK.

68. Schrire2005 (Book) : Schrire, B. D. et al (2005),', ISBN: 8773043044

69. Wikipedia (Web): Wikipedia entry, Accessdate=2010-11-14

70. Hawaii Botany (Web): University of Hawaii Botany Dept Flowering Plants Index (May be copyrighted), Accessdate=2010-11-14

71. Judd, W. S., C. S. Campbell, E. A. Kellogg, P. F. Stevens, and M. J. Donoghue. 2002. Plant Systematics, A Phylogenetic Approach, 2nd edition. Sinauer Associates, Sunderland, Massachusetts, USA.

www.ingramcontent.com/pod-product-compliance
Lightning Source LLC
Chambersburg PA
CBHW040250220526
45473CB00001B/437